EARTH UNDER CONSTRUCTION!
PLATE TECTONICS
RESHAPE EARTH!

BY ABBY BADACH DOYLE

Gareth Stevens
PUBLISHING

Please visit our website, www.garethstevens.com. For a free color catalog of all our high-quality books, call toll free 1-800-542-2595 or fax 1-877-542-2596.

Cataloging-in-Publication Data

Names: Doyle, Abby Badach.
Title: Plate tectonics reshape Earth! / Abby Badach Doyle.
Description: New York : Gareth Stevens Publishing, 2021. | Series: Earth under construction! | Includes glossary and index.
Identifiers: ISBN 9781538258224 (pbk.) | ISBN 9781538258248 (library bound) | ISBN 9781538258231 (6 pack)
Subjects: LCSH: Plate tectonics–Juvenile literature. | Faults (Geology)–Juvenile literature. | Earth sciences–Juvenile literature. | Earth (Planet)–Internal structure–Juvenile literature.
Classification: LCC QE511.4 D69 2021 | DDC 551.1'36–dc23

First Edition

Published in 2021 by
Gareth Stevens Publishing
111 East 14th Street, Suite 349
New York, NY 10003

Copyright © 2021 Gareth Stevens Publishing

Designer: Sarah Liddell
Editor: Kate Mikoley

Photo credits: Cover, p. 1 RS Smith Photography/Shutterstock.com; space background and Earth image used throughout Aphelleon/Shutterstock.com; caution tape used throughout Red sun design/Shutterstock.com; p. 5 ANDRZEJ WOJCICKI/SCIENCE PHOTO LIBRARY/Science Photo Library/Getty Images; p. 7 MARK GARLICK/SCIENCE PHOTO LIBRARY/Science Photo Library/Getty Images; p. 9 Rost9/Shuterstock.com; p. 11 Peter Hermes Furian/Shutterstock.com; p. 13 aaaaimages/Moment/Getty Images; pp. 15 (continental-continental and continental-oceanic)), 19 (diagram), 21 (diagram) Domdomegg/Wikimedia Commons; p. 15 (oceanic-oceanic) Chandres/Wikimedia Commons; p. 17 Bloomberg Creative Photos/Bloomberg Creative Photos/Getty Images; p. 19 (Krafla) Patrick Dieudonne/robertharding/Getty Images Plus/Getty Images; p. 21 (San Andreas Fault) Lloyd Cluff/Corbis Documentary/Getty Images; p. 23 Photography by ZhangXun/Moment/Getty Images; p. 25 Yarr65/Shutterstock.com; p. 27 piskunov/E+/Getty Images; p. 29 MichaelTaylor/Shutterstock.com.

All rights reserved. No part of this book may be reproduced in any form without permission in writing from the publisher, except by a reviewer.

Printed in the United States of America

Some of the images in this book illustrate individuals who are models. The depictions do not imply actual situations or events.

CPSIA compliance information: Batch #CS20GS: For further information contact Gareth Stevens, New York, New York at 1-800-542-2595.

Find us on 󰈎 󰋾

CONTENTS

Always on the Move 4

Like a Puzzle 6

One Earth, Many Layers. 8

How Many Plates Are There?. 10

Two Kinds of Crust 12

When Plates Meet...and Crash! 14

When Plates Move Apart 18

When Plates Slide Sideways 20

Three's a Crowd! 22

The Ring of Fire 24

Another Pangaea? 28

Glossary. 30

For More Information. 31

Index 32

Words in the glossary appear in **bold** type the first time they are used in the text.

ALWAYS ON THE MOVE

Take a deep breath and try to sit as motionless as you can. Surprise! You're still slowly moving, even if you can't feel it. That's because Earth's crust is slowly moving beneath your feet.

Earth's crust isn't one big piece. It's broken up into huge chunks called tectonic plates. The plates split deep below the continents and oceans. There, they float slowly on melted rock. Over millions of years, tectonic plates move to make huge mountain ranges and deep **trenches**. Their movement also causes **earthquakes** and **volcanoes**. Tectonic plates reshape Earth in a big way!

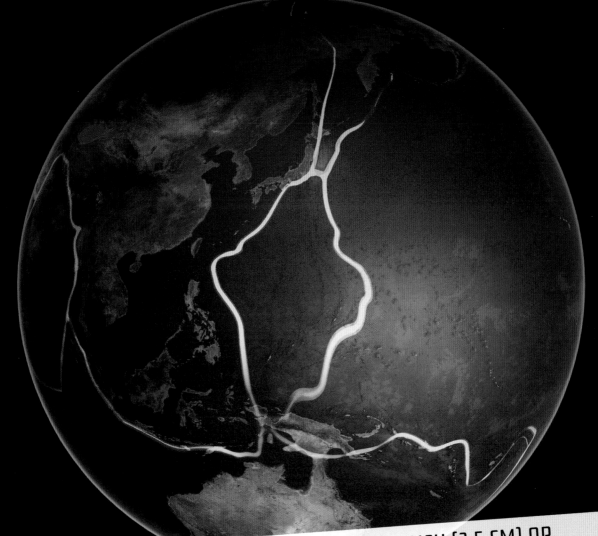

TECTONIC PLATES MOVE AS LITTLE AS 1 INCH (2.5 CM) OR AS MUCH AS 4 INCHES (10 CM) PER YEAR. ON AVERAGE, THEY MOVE AT ABOUT THE RATE YOUR FINGERNAILS GROW.

A HIGH-TECH LOOK

Geologists, or scientists that study the physical forms of Earth, use global positioning system (GPS) **satellites** to track the movement of tectonic plates. These satellites don't move in relation to Earth's crust, like something sitting on the ground would do. This gives scientists a very exact way to measure how far plates move.

LIKE A PUZZLE

Have you ever noticed that Earth's continents look like they could fit together like a puzzle? That's because, millions of years ago, they did! Scientists call this large mass of land Pangaea (also spelled Pangea).

One of the first people to come up with this theory, or scientific idea, was the German scientist Alfred Wegener. He noticed that fossils and rock patterns from continents very far apart seemed to match up. In 1912, he released his theory, which he called continental drift. He believed the continents must have traveled over time, but he wasn't sure how. Today, we know the "how" is tectonic plates.

SCIENTISTS THINK PANGAEA FIRST FORMED AS EARLY AS AROUND 280 MILLION YEARS AGO. IT STAYED THAT WAY FOR AT LEAST 50 MILLION YEARS BEFORE IT SLOWLY BEGAN TO SPLIT APART.

FOSSILS GIVE US CLUES

Fossils from *Mesosaurus*, similar to a crocodile, have been found in South Africa and Eastern South America. *Mesosaurus* could swim short distances, but not thousands of miles across the Atlantic Ocean! If the continents were once connected, it would make more sense for similar fossils to be found in both places.

ONE EARTH, MANY LAYERS

Earth is made up of separate layers. The top layer is the crust. Under Earth's oceans, the crust is between about 3 to 5 miles (5 to 8 km) thick. However, the crust that makes up the land we live on is much thicker, ranging from about 5 miles (8 km) to 43.5 miles (70 km) thick!

Under the crust is the mantle. Together, the crust and the stiff outer part of the mantle form the tectonic plates. The inner mantle is made up of very hot, heavy, liquid rock. The warmer parts move up and the colder parts move down in a big circle. This motion is what causes the tectonic plates to move.

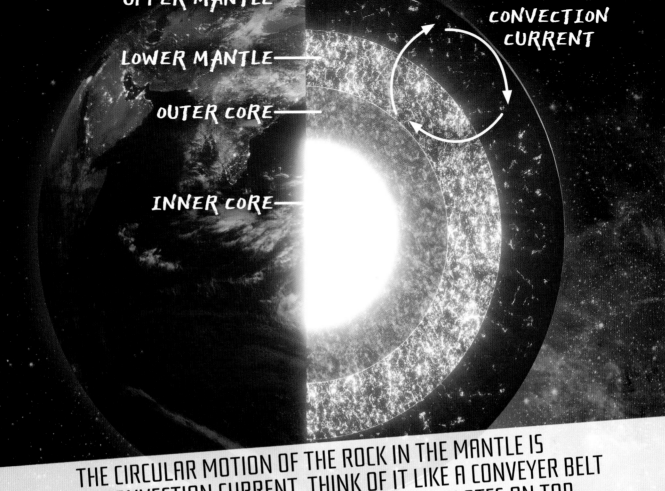

UPPER MANTLE
LOWER MANTLE
OUTER CORE
INNER CORE
CONVECTION CURRENT

THE CIRCULAR MOTION OF THE ROCK IN THE MANTLE IS CALLED A CONVECTION CURRENT. THINK OF IT LIKE A CONVEYER BELT THAT SLOWLY GLIDES, MOVING THE TECTONIC PLATES ON TOP.

LIQUID ROCK

Have you ever seen a construction crew use **asphalt** to pave a road or driveway? It's very hot and flows like a liquid. This is what scientists think rock in Earth's mantle is like. Liquid rock is about 1,600°F (871°C) at the top of the mantle. It can heat up to 4,000°F (2,204°C) closer to the core!

HOW MANY PLATES ARE THERE?

Tectonic plates exist too deep in Earth's crust to see with your eyes. However, Earth gives us clues, such as mountain ranges and volcanoes, to show where the edges, or boundaries, of each plate might be. This also shows which direction they might be moving.

Since we can't see them to count them, scientists have different opinions on exactly how many plates there are. By learning to read Earth's clues, they can make an educated guess. Most scientists agree Earth has between 15 and 24 plates. The 7 to 10 largest are called major plates. The rest are called minor plates.

EARTH'S TECTONIC PLATES

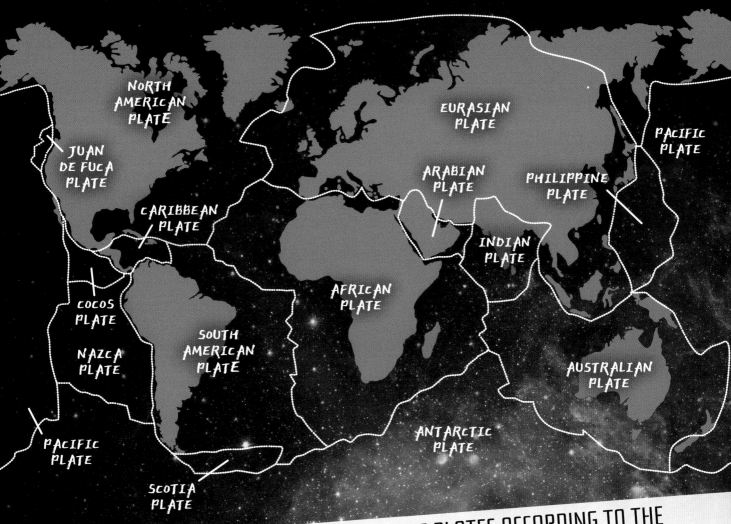

THIS MAP SHOWS THE EARTH'S TECTONIC PLATES ACCORDING TO THE UNITED STATES GEOLOGICAL SURVEY, OR USGS. MOST PLATES ARE NAMED FOR THE CONTINENT OR OCEAN THAT LIES ABOVE THEM.

LARGEST AND SMALLEST

The largest plate is the Pacific Plate, underneath the Pacific Ocean. Scientists think it is about 40 million square miles (103 million sq km) in area. One of the smallest is the Juan de Fuca Plate, at about 96,500 square miles (250,000 sq km). Though small, it causes a lot of activity.

TWO KINDS OF CRUST

Scientists put tectonic plates into two main categories, or groups: oceanic and continental. Like their names suggest, oceanic plates exist underneath oceans and continental plates are underneath land. Oceanic plates are thinner and made up of rocks rich in the **minerals** iron and magnesium. This type of rock is dense and heavy. That's why oceanic plates sink below Earth's ocean water.

Continental plates are thicker. They're made up of rock too, but it's lighter and less dense. That's why the continents sit above the water. Continental plates are made up of felsic rock. These are rocks, such as granite, that have the mineral feldspar.

OFF THE WASHINGTON COAST, THE OCEANIC JUAN DE FUCA PLATE **COLLIDES** WITH THE CONTINENTAL NORTH AMERICAN PLATE. THIS CAUSED MANY VOLCANOES TO FORM IN THE NEARBY CASCADE RANGE.

ALMOST ALL WET

The land on all Earth's seven continents adds up to about 57 million square miles (148 million square km). That's a lot! Still, most of our planet is covered by water. The area of the ocean is more than twice as big as all the continents put together. The ocean covers about three-fourths of Earth's surface.

WHEN PLATES MEET... AND CRASH!

When tectonic plates meet, they form what's called a plate boundary. There are three different types of plate boundaries. The first is called a convergent boundary. In a convergent boundary, plates move toward each other and collide. Think of it as a very slow crash that happens over millions of years.

Lots of things, like mountain ranges and volcanoes, can spring up near a convergent boundary. Exactly what happens depends on the types of plates that collide. There are three different types of convergent boundaries: between two oceanic plates, two continental plates, or one of each.

THREE TYPES OF CONVERGENT BOUNDARIES

OCEANIC-OCEANIC

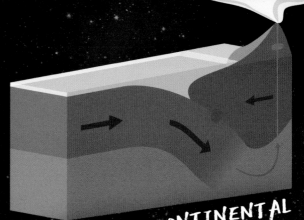

OCEANIC-CONTINENTAL

CONTINENTAL-CONTINENTAL

AT A CONVERGENT PLATE BOUNDARY, PART OF EARTH'S SURFACE IS DESTROYED. THIS TYPE OF BOUNDARY IS ALSO CALLED A DESTRUCTIVE PLATE BOUNDARY. DESTRUCT MEANS TO WRECK OR CAUSE HARM.

THE ROCK CYCLE

Rocks change from one shape to another. Weather can break down a cliff to dust. Over time, pressure can turn dust into solid rock. A convergent boundary can push rock down into the mantle to melt it into magma. Nearby volcanoes erupt **molten** rock, which cools and hardens. Then, the cycle begins again!

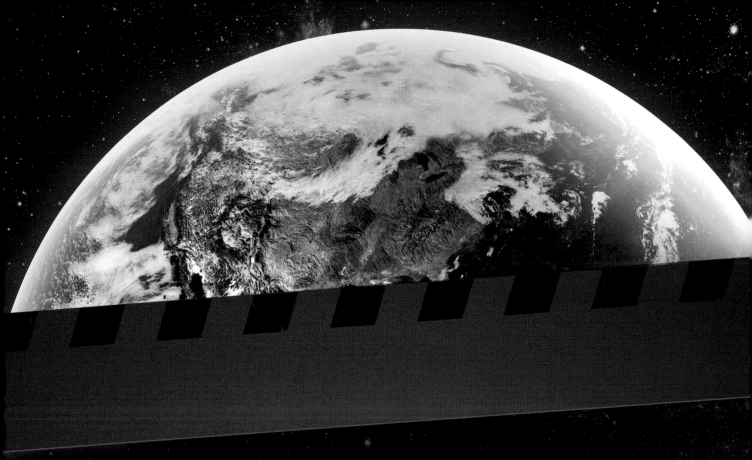

When heavy ocean crust meets another plate, the ocean crust wedges underneath it. The force pushes the ocean crust downward, where it melts into magma. Think of it like Earth's crust getting recycled! This is called subduction. Subduction can happen between two oceanic plates, or one of each type. Subduction creates deep ocean trenches. As magma rises up, it can **trigger** a volcano. Earthquakes happen from the strain of these plates interacting too.

Continental plates are too light to subduct. When these plates meet, they push up or sideways. This is how many mountain ranges, such as the Himalayas and the Appalachian Mountains, were formed.

THE HIMALAYAN MOUNTAIN RANGE WAS FORMED BY THE COLLISION OF THE INDIAN AND EURASIAN PLATES. IT IS HOME TO THE WORLD'S HIGHEST POINT, MOUNT EVEREST, AT ABOUT 29,035 FEET (8,850 M).

HIGHS AND LOWS

Convergent plate boundaries created Earth's highest and lowest points. Deep in the ocean, the Mariana Trench sits where the Pacific Plate converges with the Philippine Plate. Its deepest point is 36,201 feet (11,034 m). If you put Mount Everest on the bottom, the top of the mountain would still be more than 7,000 feet (2,134 m) underwater!

WHEN PLATES MOVE APART

The second type of boundary, a divergent boundary, happens when tectonic plates spread apart. Here, magma bubbles up through the gap and hardens. This becomes new crust.

Divergent boundaries form areas called rift valleys. Underwater, they're known as mid-ocean ridges. The Mid-Atlantic Ridge is an underwater mountain range that extends below the Atlantic for 10,000 miles (16,093 km) from the Arctic Ocean to the southern tip of Africa. Its valleys are as wide as 9 miles (14.5 km), formed by millions of years of slowly splitting apart. Scientists think it widens by about 1 inch (2.5 cm) per year.

KRAFLA IS ONE OF THE MOST ACTIVE VOLCANIC SYSTEMS IN ICELAND. IT IS LOCATED RIGHT ON TOP OF THE DIVERGENT BOUNDARY OF THE NORTH AMERICAN AND EURASIAN CONTINENTAL PLATES.

UNDERWATER HOT SPRINGS

Divergent boundaries underwater can also release jets of hot water, called chimneys or hydrothermal vents. The East Pacific Rise and the Galápagos Rift are known for these. Weird creatures like tube worms, vent crabs, and giant clams **thrive** nearby. Some fish use the chimneys to keep their eggs warm!

WHEN PLATES SLIDE SIDEWAYS

When plates move past each other **horizontally**, it's called a transform boundary. The word "transform" means to change something. At a transform boundary, Earth's crust isn't created or destroyed. However, it does change shape. . .and cause trouble! Transform boundaries are known for causing earthquakes.

Most transform boundaries happen underwater. They form in a zigzag pattern and leave marks on the ocean floor. Sometimes they form on land. The San Andreas Fault, a transform boundary in California, is one famous example. The action along this boundary caused one of the most destructive earthquakes of all time in San Francisco, California in 1906.

THE SAN ANDREAS FAULT EXTENDS FOR 808 MILES (1,300 KM) THROUGH CALIFORNIA. IT MOVES AN AVERAGE OF 2 INCHES (5 CM) PER YEAR.

IT'S ALL YOUR FAULT!

At the San Andreas fault, the fast-moving Pacific Plate slowly grinds against the North American Plate. This action has occurred for about 10 million years. Here, the Pacific Plate moves in a slight northwest direction, pointing toward Alaska. The North American Plate moves slightly southeast, pointing toward Mexico.

THREE'S A CROWD!

When three plates meet, it's known as a triple junction. The word "junction" means a place where things join together. Three is the highest number of plates that can come together. Four or more plates almost never meet. When they do, they only stay for an instant. One notable triple junction is the Afar triple junction, near the area between the continent of Africa and the country of Saudi Arabia.

Scientists put triple junctions into two categories: stable and unstable. Stable means they can stay in that arrangement over time. Unstable means they can only stay for a moment before they move to a different position.

MOUNT FUJI IS AN ACTIVE VOLCANO AND JAPAN'S TALLEST MOUNTAIN. THREE TECTONIC PLATES MEET NEAR HERE: THE AMUR PLATE, OKHOTSK PLATE, AND PHILIPPINE PLATE.

ALPHABET SOUP

Geologists use letters to quickly describe the boundaries at triple junctions. The letters stand for the features that form there: R for ridge, or divergent; T for trench, or convergent; and F for fault, or transform. For example, an RFF triple junction has one divergent and two transform boundaries.

THE RING OF FIRE

The world's most famous region of tectonic activity is the Ring of Fire. It gets this nickname from more than 450 volcanoes that circle it. In fact, about 75 percent of all Earth's volcanoes are found here—many of them underwater! Many of Earth's deepest ocean trenches, including the Mariana Trench, are along the Ring of Fire.

Nearly 90 percent of earthquakes happen here too. The Ring of Fire extends for 24,900 miles (40,073 km) around the Pacific Plate. It actually is shaped more like a horseshoe, but the name has a nice "ring" to it, don't you think?

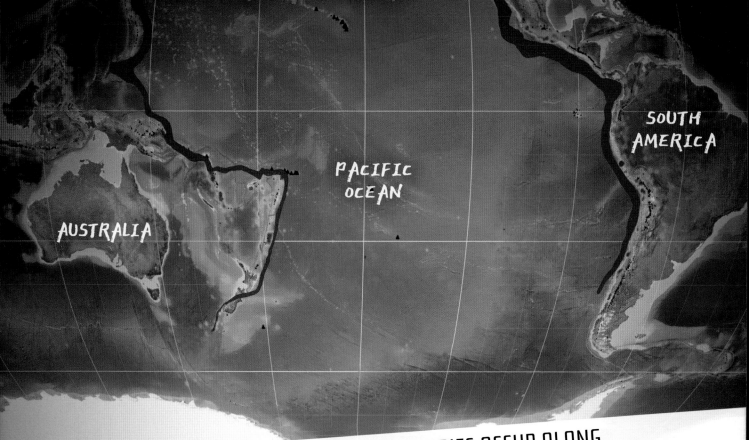

ALL THREE TYPES OF PLATE BOUNDARIES OCCUR ALONG THE RING OF FIRE: CONVERGENT, DIVERGENT, AND TRANSFORM. MANY TRIPLE JUNCTIONS EXIST HERE TOO.

VOLCANIC ISLANDS

The movement of tectonic plates can sometimes cause islands to form. The Aleutian Islands, located off the coast of mainland Alaska, are along the northern arc of the Ring of Fire. Numerous volcanoes exist along the more than 300 individual islands. They were created when the Pacific Plate subducted beneath the North American Plate.

Many large and small tectonic plates circle the Pacific Plate along the Ring of Fire. The major plates in the east include the North American, South American, and Nazca Plates. To the south is the Antarctic Plate. The biggest plates along the west include the Indian-Australian, Philippine, and Eurasian Plates.

Finally, the tiny Juan de Fuca Plate sits along its northeast curve. It might be small, but subduction zones here cause a lot of tectonic activity. One notable event was the eruption of the Mount Saint Helens volcano on May 18, 1980. This triggered the largest landslide in recorded history!

NEW ZEALAND'S MOUNT RUAPEHU IS LOCATED ON THE RING OF FIRE WHERE THE PACIFIC AND AUSTRALIAN PLATES MEET. IT HAS A SMALL ERUPTION YEARLY AND A LARGE ONE ABOUT EVERY 50 YEARS.

UNDERWATER VOLCANOES

Scientists have only just begun to explore the underwater depths of the Ring of Fire. To study the deep trenches, scientists need special submarines, or underwater ships, that can handle the **intense** pressure of the deep ocean. Hundreds of volcanoes exist here. When they erupt, it can trigger earthquakes, landslides, and **tsunamis**.

ANOTHER PANGAEA?

Scientists don't know for sure what the future holds. Earth is full of surprises! However, they do know how fast the tectonic plates are moving and in what direction. Using that data, scientists can paint a picture of what the Earth might look like in millions of years.

The result looks familiar: all the continents packed together in one big landmass. It's Pangaea all over again! Scientists call this someday landmass Pangaea Proxima or Pangaea Ultima, for the Latin words meaning "next" or "last." Generations from now, humans might be able to drive a car from Australia to Antarctica—but only time will tell.

THIS IMAGE SHOWS WHAT EARTH COULD LOOK LIKE IN 250 MILLION YEARS. THE SMALL SEA IN THE MIDDLE IS ALL THAT MIGHT REMAIN OF THE ATLANTIC OCEAN.

ON THE MOVE

Some scientists think Africa will continue to move north and one day collide with Europe. That movement could close up the Mediterranean Sea into a mountain range, just like what happened to the Himalayas. North and South America may become much closer together too.

GLOSSARY

asphalt: matter often used to make roads

collide: the hitting of two objects against one another

earthquake: a shaking of the ground caused by the movement of Earth's crust

horizontally: in a manner level with the line that seems to form where Earth meets the sky

intense: existing in an extreme degree

mineral: matter in the ground that forms rocks

molten: changed into a liquid form by heat

satellite: an object that circles Earth in order to collect and send information or aid in communication

thrive: to grow or develop successfully

trench: a long, narrow hole in Earth's crust

trigger: to cause something to start or happen

tsunami: a huge wave of water created by an underwater earthquake or volcano

volcano: an opening in a planet's surface through which hot, liquid rock sometimes flows

FOR MORE INFORMATION

BOOKS

Greer, Eileen. *Plate Tectonics*. New York, NY: PowerKids Press, 2017.

London, Martha. *Looking Inside Earth*. North Mankato, MN: The Child's World, 2020.

Quinlan, Julia J. *Plate Tectonics*. New York, NY: Britannica Educational Publishing, 2018.

WEBSITES

BrainPOP: Plate Tectonics
www.brainpop.com/science/earthsystem/platetectonics/
This interactive site has cool maps, worksheets, a quiz, and more fun ways to learn about plate tectonics.

NASA Space Place: Tectonic Forces
spaceplace.nasa.gov/tectonics-snap/en/
Watch a short video or download a poster to learn how tectonic plates move.

PBS: Curious Crew Plate Tectonics
www.pbs.org/video/curious-crew-plate-tectonics-ep-208/
Tag along with the Curious Crew as they do a science project about earthquakes and learn about Earth's layers.

Publisher's note to educators and parents: Our editors have carefully reviewed these websites to ensure that they are suitable for students. Many websites change frequently, however, and we cannot guarantee that a site's future contents will continue to meet our high standards of quality and educational value. Be advised that students should be closely supervised whenever they access the internet.

INDEX

Aleutian Islands 25

continental plates 12, 14, 16, 19

continents 4, 6, 7, 11, 12, 13, 14, 15, 16, 19, 22, 28

convection current 9

convergent boundary 14, 15, 17, 23, 25

divergent boundary 18, 19, 23, 25

fossils 6, 7

geologists 5, 23

hydrothermal vents 19

Juan de Fuca Plate 11, 13, 26

magma 15, 16, 18

mantle 8, 9, 15

Mid-Atlantic Ridge 18

Mount Everest 17

Mount Fuji 23

Mount Ruapehu 27

Mount Saint Helens 26

North American Plate 13, 21, 25

oceanic plates 12, 14, 16

Pacific Plate 11, 17, 21, 24, 25, 26

Pangaea (Pangea) 6, 7, 28

Ring of Fire 24, 25, 26, 27

San Andreas Fault 20, 21

subduction 16, 26

transform boundary 20, 23, 25

triple junction 22, 23, 25

Wegener, Alfred 6